T0193447

Cameron Loves Cats and Cameron Loves to Count

Dr. Edward G. Smith, M. Ed., Ph. D.

Archway Publishing books may be ordered through booksellers or by contacting:

Archway Publishing
1663 Liberty Drive
Bloomington, IN 47403
www.archwaypublishing.com
844-669-3957

Because of the dynamic nature of the Internet, any web addresses or links contained in this book may have changed since publication and may no longer be valid. The views expressed in this work are solely those of the author and do not necessarily reflect the views of the publisher, and the publisher hereby disclaims any responsibility for them.

Any people depicted in stock imagery provided by Getty Images are models, and such images are being used for illustrative purposes only.
Certain stock imagery © Getty Images.

Interior Image Credit: Dr. Edward G. Smith, M.Ed., Ph.D.

ISBN: 978-1-6657-2772-3 (sc)
ISBN: 978-1-6657-2773-0 (e)

Print information available on the last page.

Archway Publishing rev. date: 09/28/2023

Cameron Loves Cats and Cameron Loves to Count

Cameron Loves Cats
She loves their eyes
and the way they look about
She loves their mouths
and the way they meow
She loves their legs
and the way they pounce
But most of all
Cameron loves to count

One Cat

Two eyes-Look about
One mouth- Meow
Four legs- Pounce

Two Cats

Four eyes- Look about, Look about
Two mouths- Meow, Meow
Eight legs- Pounce, Pounce

Three Cats

Six eyes- Look about, Look about, Look about
Three mouths- Meow, Meow-Meow
Twelve legs- Pounce, Pounce, Pounce

Cameron Loves Cats
Calico, Siamese
Some like milk
Some like cheese

Some short and fat
Some tall and thin
Some with furry feet
Some with just skin

Some very high
Some very low
And if they are about,
you may not know

But most of all
Cameron Loves to count

Four Cats

Eight eyes- Look about, Look about, Look about, Look about
Four mouths- Meow, Meow, Meow, Meow
Sixteen legs- Pounce, Pounce, Pounce, Pounce

Five Cats

Ten eyes- Look about,
Look about, Look about,
Look about, Look about
Five mouths- Meow, Meow,
Meow, Meow, Meow
Twenty legs- Pounce, Pounce,
Pounce, Pounce, Pounce

Cameron Loves Cats
She loves their sighs
She loves their eyes
Some close together
Some wide apart
Some even glow in the dark
But most of all
Cameron Loves to Count

Printed in the United States
by Baker & Taylor Publisher Services